Frank Brinkley, Kakuzo Okakura

Japan

Described and illustrated by the Japanese

Frank Brinkley, Kakuzo Okakura

Japan
Described and illustrated by the Japanese

ISBN/EAN: 9783741175282

Manufactured in Europe, USA, Canada, Australia, Japa

Cover: Foto ©ninafisch / pixelio.de

Manufactured and distributed by brebook publishing software
(www.brebook.com)

Frank Brinkley, Kakuzo Okakura

Japan

JAPAN ○

SECTION
X

JAPAN

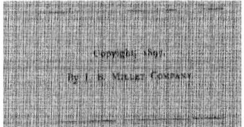

THE ATTITUDE OF JAPAN TOWARD FOREIGN RESIDENTS.

JAPANESE FINANCE.

(Concluded.)

E have dwelt upon this chapter of Japan's modern history at some length, not merely because it sets forth a fine feat of finance, indicating clear insight, good organizing capacity and courageous energy, but also because volumes of adverse foreign criticism were written into its margin during the course of the incidents it embodies. A score of onlooking strangers were prepared each with an infallible nostrum of his own, the rejection of which convinced him of Japan's hopeless stupidity. Now she was charged with robbing her own people because she bought their goods with paper money and sold them for specie; again she was accused of an official conspiracy to ruin the foreign local banks because she purchased exporters' bills on Europe and America at rates that defied ordinary competition; and while some declared that she was plainly without any understanding of her own doings, others predicted that she could not possibly extricate herself from the slough of an inflated and largely depreciated fiat currency without recourse to European capital, and agreed that her heroic method of dealing with the problem would paralyze industry, interrupt trade, produce widespread suffering, and, in short, bring about the advent of the proverbial seven devils. Undoubtedly, to carry the currency of a nation from a discount of seventy or eighty per cent to par in the course of four years, reducing its volume at the same time from one hundred and thirty-five to one hundred and nineteen millions, was a financial enterprise violent and daring almost to rashness. The gentler expedient of a foreign loan—an expedient of recently proved efficacy in Italy's case—would have commended itself to the majority of economists. But it may be here stated, once for all, that until her adoption of gold monometallism in 1897, the foreign money market was practically closed to Japan. Had she borrowed abroad, it must have been on a sterling basis. Receiving a fixed sum in silver, she would have had to discharge her debt in rapidly appreciating gold. Twice, indeed, she had recourse to London for small sums, but when she came to cast up her accounts, the cost of the accommodation stood out in deterrent proportions.[1] These considerations were supplemented by a strong aversion to incurring pecuniary obligations to Western states before the latter consented to restore her judicial and tariff autonomy. The example of Egypt showed what kind of fate might overtake a semi-independent state falling into the clutches of foreign bondholders. Japan did not wish to fetter herself with foreign debts while struggling to emerge from the ranks of Oriental powers. After all, nothing succeeds like success. Japanese financiers made a signal success. Having undertaken to reorganize the administration of an empire and inaugurate a vast programme of reform, they met the difficulty of an empty treasury by issuing fiat notes, and then, fourteen years later, grappling boldly with the problem of this inflated and heavily depreciated currency, they restored its value to par and resumed specie payments in the brief space of four years. This brings us back to the point at which we digressed to speak of the currency question.

The volume of the foreign trade grew from twenty-six million yen in 1868 to forty-nine millions in 1873, and developed so slowly during the next decade that, after vibrating between the latter figure and sixty-seven millions, it became almost stationary, so that the public began to despair of any large growth, and the fair prospects of the early days faded out of sight. Yet there was good reason to wonder that trade could continue at all under the circumstances. Against the import merchant the currency trouble worked with double potency. Not only did the gold with which he purchased goods appreciate constantly in terms of the silver for which he sold them, but the silver itself appreciated sharply and rapidly in terms of the fiat notes paid by Japanese consumers. Cursory reflection may suggest that these factors should have operated inversely to stimulate exports as much as they depressed imports. But such was not altogether the case in practice. For the exporters' transactions were always hampered by the possibility that a delay of a week or even a day might increase the purchasing power of his silver by bringing about a further depreciation of paper, and it was not till this element of pernicious disturbance was removed that the trade recovered a healthy tone and grew so lustily that in 1897 its volume aggregated three hundred and eighty-two million yen, thus treading close on the heels of the foreign commerce of China, with her three hundred million inhabitants and long-established international relations.

Two questions of prime interest suggest themselves here: first, how long will this trade remain in the hands of foreign middlemen? secondly, what are the prospects of its future? As to the former point, statistics suggest that the Japanese are steadily pushing out the foreigner. Looking at the past decade, we find that whereas, in 1888, Japanese merchants carried on only twelve per cent of the total trade without the intervention of foreign middlemen, their share rose to thirty-two per cent in 1897. It is natural, of course, that an energetic

[1] The silver coin then placed on the Timolon market in 1868 and sold at 87½ cents produced 4.78 ounces, yet one, altogether, 1,1869.575,972 yen; a commencement that level in 1874 and sold at 61 cents produced 10,633,000 yen, altogether, 37,000,000 yen, respectively.

effort should be made by the people of the country to carry on their own commerce independ
ently of alien assistance. But some special features of the foreigner's methods in Japan render
his intervention particularly irksome. Thus, in purchasing raw silk, his habit is to take the
staple into his warehouse and inspect it there at his leisure before completing the bargain;
and in the case of tea he
buys the leaf in parcels
without discrimination
as to their *provenance*,
dumps them all together
into his firing-pans, and
packs the refired and
recolored article for ex-
port in boxes bearing
his own *cachet*. That
the former method was
originally necessitated
by the impossibility of
trusting the Japanese to
sell silk honestly by
sample, and the latter

REELING SILK.

The cocoons are soaked in hot water while unwinding. Live coals under the iron pot keep the water at a heat
that frequently parboils the operator's hands.

by their want of skill to prepare the tea for Western markets, is well understood. But if these
reasons justify the foreigner's procedure, they are certainly not of a nature to appease the
native's sensitiveness. The ambition of the Japanese to displace the foreign middleman must
grow. Its complete gratification will be long postponed, however. The foreign resident merchant is an ideal agent. As an exporter his command of cheap capital, his experience, his
knowledge of foreign markets and his connections enable him to secure prices which Japanese, working on their own account, could not obtain; as an importer he enjoys credit abroad
which the Japanese are without, he pays to Japanese producers ready cash for their staples,
taking upon his own shoulders all the risks of finding a sale for them beyond the sea, and he
offers to Japanese consumers imports laid at their doors without any responsibility on their
own part. Further, direct dealings between foreign merchants in Europe or America and
Japanese merchants in Japan could not be undertaken with safety to the former. The asser-
tion sounds harsh, but in truth Japanese traders have not yet developed the commercial conscience which is the basis of all sound business. Exceptions to this rule are numerous, of
course, and their number grows steadily. But it has to be recorded, with regard, at any rate,
to the Japanese coming into tradal contact with foreigners, that neither the moral sanctity of
an engagement nor the material advantage of credit and confidence, nor even the practical
necessity of implementing every condition of a contract, is fully appreciated by the average
man of affairs.

In China there are guilds whose chief object is to strengthen credit. Lest the business of the members in general should lose the benefit of public trust, they make good the default of any one of their number. In Japan also there are guilds, but their disposition is to shield and abet the defaulter rather than to discountenance him when a foreigner is his victim. The causes chiefly responsible for this state of affairs are not difficult to analyze. One fact, constantly adduced and certainly deserving prominence, is that the Japanese frequenting the treaty ports and doing business with the foreign resident belong to a distinctly inferior stratum of the nation. They established their footing at a time when all contact with foreigners was counted degrading or unpatriotic. For the most part they were men without reputations to imperil, and they approached the foreigner with a disposition to regard him as a person to be neither spared nor respected. In short, they were not, nor are they yet, fair representatives of the upper grade of Japanese merchants. A more subtle factor is that the wholesome atmosphere of public opinion is virtually wanting in the region of this open-port trade. Whatever chicanery a Japanese may practise against foreigners, his own version of the incident alone reaches his nationals. Opinions may differ as to the efficacy of the checks which the scrutiny of his fellows imposes upon the average mortal's improbity, but that it does impose a considerable check, none will deny. The Japanese in his dealings with foreign resident merchants is beyond the influence of such checks. If he sins, it is with the comfortable conviction that

his sin will not find him out. Finally, there is the broad fact that from early times trade stood in the lowest rank of Japanese bread-winning occupations. The trader was not respected and did not respect himself. That prejudice, peculiar to a military society, has now disappeared in theory, but its practical consequences cannot be easily effaced. There is, indeed, no

SORTING AND SIFTING TEA OUTSIDE THE GO-DOWN (STOREHOUSE).

warrant for attributing moral deficiency to the Japanese race. If their standard of truth loses something by comparison with ours in the abstract, there is a balance of practical ingenuousness in their favor. The falsehoods covertly sanctioned by the conventions of social intercourse and every-day life in the Occident are openly permitted in Japan. Truth derives value less from its independent nobility than from the nature of its consequences. To tell the truth where to withhold or even to transform it would avert misfortune greater than the moral penalty attaching to all subterfuge, is not Japanese philosophy, any more than to obtrude personal griefs upon the notice of those whom they do not concern is a canon of Japanese courtesy. Sorrow paraded in public is either a selfish exaction of sympathy or an insincere desire to be credited with profound feeling. The truth spoken without regard for results is either the prompting of giddiness or a bid for the reputation of personal integrity at the expense of other people's happiness. That is the acknowledged creed of Japan; the partially practised but unconfessed creed of the Occident also. But for the rest, the fibre of the Japanese conscience seems to be just as tough as the fibre of any other conscience, and not more elastic. Commercial morality, however, which is the special outgrowth of trading habits, is still a stunted plant in Japan, and until it attains much larger proportions the foreign middleman will be an indispensable figure in the country's international commerce.

We come now to the question, in what directions may the country's commerce be expected to expand; or, speaking in broader terms, what may be regarded as the wealth-earning capacities of the Japan of the future? For the purposes of such an inquiry, the first point to be determined is whether the development will be in the production of raw materials or of manufactured goods. The answer admits of no doubt. Japan will always be able to send abroad considerable quantities of silk and tea, and inconsiderable quantities of marine products,[1] copper, coal,[2] camphor, sulphur, rice and minor staples; but with regard to these, either her producing capacity is inelastic or her market is limited. It is certain, indeed, that she will by and by have to look abroad for supplies of the necessaries of life. Rice is the staple diet of her people, and she seems to have almost reached the potential maximum of her rice-growing area; for in spite of her genial climate and seemingly fertile soil, the extent of her arable land is disproportionately small. She has only eleven and one half millions of acres under crops, and there is no prospect of any large extension, or of the yields being improved by new agricultural processes. The Japanese farmer understands his work thoroughly. His competence is sufficiently proved when we say that by the skilful use of fertilizers he has been able to raise good crops of rice on the same land during fifteen or twenty centuries. On the other hand, not only is the population increasing rapidly, but in proportion to the growth of general prosperity and the distribution of wealth the lower classes of the people, who used formerly to be content with barley and millet, now regard rice as an essential article of food. It cannot be long, therefore, before large supplies of this cereal will

[1] Japan's fishing industry is doubtless capable of great development. She has 17,602 miles of coast and 270,000 families devoted to fishing, or more than 15 families to each mile. They employ 330,796 boats and 1,194,408 nets, representing a capital of about twenty-eight million yen, and the total value of the annual catch is put at forty-eight million yen, though one hundred millions would probably be nearer the truth. The fishermen are sturdy, courageous fellows, but their methods are primitive, and virtually no improvements have yet been introduced.

[2] It was at one time supposed that Japan possessed great mineral wealth, but experience has corrected the impression. The output of her various mines increases steadily, it is true, but its total annual value does not exceed thirty million yen.

have to be drawn from abroad. The same is true of timber, which has already become inconveniently scarce. Japan cannot even grow her own cotton, and nature has not fitted her pastures for sheep; so that materials for her people's clothing have all to be imported. Her future lies undoubtedly in industrial enterprise. She has an abundance of cheap labor and her people are exceptionally gifted with intelligence, docility, manual dexterity and artistic taste. Everything points to a great future for them as manufacturers. This is not a matter of mere conjecture. Striking practical evidence has already been furnished. Cotton spinning may be specially referred to. As long ago as 1862, the feudal chief of Satsuma started a mill with five thousand spindles in his fief, but during a whole decade he found only one imitator. In 1882, however, a year which may be regarded as the opening of Japan's industrial era, this enterprise began to attract capital, and in the course of four years fifteen mills were established working fifty-five thousand spindles. By foreign observers this new departure was regarded with contemptuous amusement. The Japanese were declared to be without organizing capacity, incapable of sustained energy and generally unfitted for factory work. These desponding views had soon to be radically modified, for by 1897 the number of mills had increased to sixty-three, the number of spindles to some eight hundred thousand, the capital invested to twenty-one million *yen*, and the average annual profit per spindle was three and one half *yen*, or thirteen and one third per cent on the capital. The rapidity of this development suggests unsoundness, but speed is a marked characteristic of Japan's modern progress. In 1880, for example, a man named Isozaki of Okayama prefecture carried to Kobe a specimen of a new kind of floor-mat, the outcome of two years' thought and trial. Briefly described, it was matting with a weft of fine green reeds and a warp of cotton yarn, having a colored design woven into it. Isozaki found difficulty in getting any one to test the salability of his invention by sending it abroad. Sixteen years later, the "brocade matting" industry of Okayama prefecture alone occupied seven hundred and thirty-four weaving establishments with nine thousand and eighty-five stands of looms, gave employment to nine thousand three hundred and fifty-seven artisans, of whom five thousand three hundred and thirty-five were females, and turned out two and one quarter million *yen* worth of this pretty floor covering. Meanwhile, the total value of the industry's output throughout the empire had reached nearly six million *yen*, and the quantity exported stood at three millions, approximately, in the customs returns. Here, then, is a trade which rose from nothing to a position of great importance in sixteen years. Even more remarkable in some respects has been the development of the textile industry. In 1886 the total production of silk and cotton fabrics was eighteen million *yen*; ten years later it had increased to ninety-six millions,[1] the number of weaving establishments

[1] The Japanese have been skilled weavers for many centuries, but a great impetus was given to this enterprise by the introduction of improved machinery and the use of aniline dyes after the opening of the country to foreign intercourse. Indigo has always been the staple dyestuff of the country. Twenty million *yen* worth is produced annually. But for colors other than blue and its various tones, aniline dyes are now imported to the extent of one and one quarter million *yen* yearly. The growth of the textile industry has also been greatly stimulated by the

being six hundred and sixty thousand four hundred and eighty, the number of looms nine hundred and thirty-nine thousand one hundred and twenty-three, and the number of operatives one million forty-two thousand eight hundred and sixty-six, of whom nine hundred and eighty-five thousand three hundred and sixteen were females. The manufacture of luci-

fer matches is another industry of entirely recent growth. A few years ago Japan used to import all the matches she needed, but by 1897 she was able not only to supply her own wants but also to send abroad five and one half million *yen* worth. Without carrying these statistics to wearisome length, we may confine ourselves to noting that in six branches of manufacturing industry which may be said to have been called into active existence by the opening of the country namely, silk and cotton fabrics, cotton yarns, matches, fancy matting and straw braid Japan's exports in 1888 aggregated only three and one quarter million *yen*, whereas the corresponding figure for 1897 was forty-two and one quarter millions. In short, the export increased thirteen hundred per cent in a decade.

STONE BRIDGE IN THE MITO GARDEN, TOKYO.

With such results before us, it is impossible to doubt that Japan has a great manufacturing future. The fact has, indeed, been partially recognized and much talked of within the past few years, especially in the United States, where the prospect of Japanese industrial competition was recently presented to the public in almost alarming proportions. On the other hand, among foreigners resident in Japan the general estimate of native manufacturing capacity is low. Doubtless, as is usually the case, the truth lies between the two extremes. Japanese industrial competition will be a formidable fact one of these days, but the time is still distant. Progress is checked by one manifest obstacle, defective integrity. Concerning every industry whose products have found a place in the catalogue of modern Japan's exports, the same story has to be told: just as really substantial development seemed

introduction of cotton yarns of fine and uniform quality. Formerly all cotton cloths were woven out of coarse, irregular, hand-spun yarns, so that nothing like regularity of weight and texture could be secured. It thus appears that Japan owes the remarkable development of her textile industry to foreign intercourse.

about to be obtained, fraudulent adulteration or dishonestly careless technique interfered to destroy credit and disgust the foreign consumer. The Japanese deny that the whole responsibility for these disastrous moral *laches* rests with them. The treaty-port middleman, they say, buys so thriftily that high-quality goods cannot be supplied to him. That excuse may be partially valid, but it is certainly not exhaustive. The vital importance of establishing and maintaining the reputation of an article offered newly in markets where it has to compete with rivals of old-established excellence is not yet fully appreciated in Japan. As to organizing capacity, the possession of which by the Japanese has been strenuously doubted by more than one foreign critic, there are proofs more weighty than any theories. In the cotton-spinning industry, for example, the Japanese are brought into direct competition in their own markets with Indian mills, employing cheap native labor, organized and managed by English-men and having the raw material at their doors. The victory rests with the Japanese, from which it may fairly be inferred that their organization is not specially defective or their method costly.[1] Yet there is one consideration that must not be lost sight of. It is the inexperience of the Japanese—their lack of standards. Japan is dressing herself in a material civilization that was made to the measure of alien nations, and curious misfits are inevitably developed in the process. If the England of 1837, for example,—that is to say, England as she was at the commencement of the Victorian era,—could have been suddenly projected forward to 1897 and invited to adapt herself to the moral and material conditions of the latter period, the task, though almost inconceivably difficult, would have been far easier than that which Japan set herself twenty-five years ago, for England would at least have possessed the preliminary training, the habit of mind and the trend of intelligence, all of which were wanting to Japan. That essential difference should be easy to remember, yet it is constantly forgotten by observers of Japan's progress. Again and again they make the mistake of measuring her acts by the standards to which they have themselves been educated. Again and again they fall into the error of deducing from her failures and perplexities the same inferences that similar perplexities and failures would suggest in Europe or America.

If the citizens of Tokyo hesitate to spend large sums upon street repair, they are accused of blind parsimony, though the fact is that, never having had any practical knowledge of really fine roadways, they have not yet learned to appreciate them. If Japanese officials do not at once succeed in solving the very difficult problem of Formosan administration, it is concluded that they lack administrative ability, though absolute lack of experience suffices to account for their ill-success. If the people have not yet made any significant contribution to the sum of Occidental scientific knowledge or mechanical contrivances, they are dismissed as imitative, not initiative; which is much as though we should charge a lad with want of originality because, having barely mastered the integral calculus, he did not write

[1] Japanese mills are kept at work twenty-three hours out of the twenty-four with one shift of operatives, and their production per spindle is forty per cent greater than the production at Bombay mills and nearly double of the production at English mills.

GREAT BELL AT CHION-IN TEMPLE, KYOTO.

This bell is the largest in Japan; it was cast in 1633 and weighs 74 tons; it is 11 feet in height, 9 feet in diameter, and nearly a foot in thickness. The deep rich tone of this bell is very musical and impressive.

some new chapters on quaternions. If they have not yet reduced constitutional government to a smoothly working system, have not yet emerged from a confusion of political coteries into the orderly condition of two great parties each capable of assuming and discharging administrative responsibilities, they are declared unfit for representative institutions, though they have tried them for only six years after fifteen centuries of military feudalism or hereditary oligarchy. If they do not carry on their new industries with the minimum of efficient labor, and if they fail to appreciate the economical necessity of bestowing constant care upon the machinery and seeking to rise above first results, instead of regarding them as the *ne plus ultra* of subsequent achievement, they are pronounced radically deficient in the industrial instinct, whereas the truth is that they have not as yet any accurate perception of the standards which experience and competition have established in foreign countries. The condition of their army and of their navy shows that not capacity but practice is what the Japanese lack. These two services are altogether modern creations. Cursory students of Japanese history rise from the perusal with a conviction that they have been reading the records of an essentially military race, and that good weapons alone are needed to make good soldiers of such people. But if the history of ancient and mediæval Japan teaches anything, its lesson is that the martial *morale* could formerly be claimed for only a very small section of the Japanese nation. The *Samurai* class numbered three millions among forty, less than one thirteenth of the total population, and to the *Samurai* class were confined the privilege of carrying swords and all the honor and distinctions attaching to that badge of aristocracy from the very beginning of the nation's social organization. The peasant, the artisan and the trader were crushed under the armed heel of the soldier, and if long centuries of enforced

MORTUARY BRONZE LANTERNS IN THE TEMPLE ENCLOSURE AT SHIBA PARK, TOKYO.

and confessed inferiority and contemptuous seclusion from camp and court have any injurious influence upon the spirit of virility and self-respect, the Japanese people, as a whole, should have been found conspicuously lacking in that spirit when the feudal system fell and the traditional distinctions of caste were abolished at the beginning of the *Meiji* era. It was, nevertheless, from the mass of the people, not from the *Samurai*, that the rank and file of the army and navy had to be taken after the Restoration. The new conscription law paid no attention to the social distinctions rigidly observed under the feudal regime. The three despised classes — the farmers, the mechanics and the merchants — found themselves suddenly

required to bear arms and to discharge duties for which they had been taught to believe themselves morally incompetent. None of the daring enterprises essayed by the makers of modern Japan attracted more interest than this reversal of the conditions which had hitherto been regarded as the bases of Japanese society — this appeal to the plebeian to enter the field specially reserved to the patrician during fifteen centuries. Could the new system work? Could

WATER-CARRIER FILLING HIS PAILS AT A WELL IN A SIDE STREET IN KYOTO.

a trustworthy *personnel* for the army and navy be obtained from such materials? While these questions were still fresh upon men's lips they were answered by a rude and conclusive test. Five years after the demise of feudalism a rebellion broke out in the south of Japan, and the regiments of conscripts had to be pitted against the very *élite* of the Japanese *Samurai*, — the two-sworded men of Satsuma. The astonishing results were that the rustic and the city clerk showed themselves almost as good fighters as the *Samurai*, and the government demonstrated that it had been able to organize an army after Western models, and that its officers could conduct a campaign in accordance with the rules of Western strategy and tactics. Seventeen years later, Japan entering the field against China furnished a conclusive proof of the excellence of her military organization. She had to undertake the most difficult task that falls to the lot of a belligerent, — the task of sending over sea two *corps d'armée* (aggregating a hundred and twenty thousand men) and maintaining them for several months in widely separated fields — one in eastern and central Manchuria, the other in the Liaotung peninsula and subsequently in Shantung province. The effort did not appear to embarrass her. There was no sign of confusion or perplexity; no breakdown of the commissariat or transport arrangements; no failure of the ambulance or hospital service. Everything worked smoothly, and the public were compelled to recognize that Japan had not only elaborated a very efficient piece of military mechanism, but also developed ability to employ it to the best advantage. The same inference was suggested by her navy. Although during two and a half centuries her people had been debarred by arbitrary legislation from navigating the high seas, the twenty-fifth year after the repeal of these crippling laws saw the state in possession of a squadron of thirty-three serviceable ships of war, officered and manned solely by the Japanese, constantly manœuvring in distant waters without accident, and evidently possessing all the qualities of a fine fighting force. In the war with China (1894-5) this navy showed its capacity by destroying or capturing, without the loss of a single ship, the whole of the enemy's fleet, whereas the latter's superiority in armor and armament ought to have produced a very different issue. It may be noted here, although the fact is foreign to our immediate subject, that for all military or naval purposes Japan possesses an immense advantage over China.

The Japanese *Samurai* is an ideal officer. Hardy, intelligent, fearless, ready to share the privations of his men and to abandon to them all the merit of victory, with little taste for luxury, a strong sense of duty and absolute devotion to his profession, he has all the qualities of a successful leader. China is entirely without such men. Even if the mass of her people could be imbued with the strong, almost fanatical, spirit of patriotism that is now known to fire the commonest Japanese, making him willing to sacrifice his life at any moment for king and country, she would still lack the traditions which in Japan exalt the profession of arms above all other occupations, and the instincts that distinguish the *Samurai* type. But whatever allowance be made on account of the splendid material furnished by the *Samurai* class for officering an army and a navy, the general fact remains that the Japa-

TEMPLE BELL AT KAWASAKI.
A village between Tokyo and Yokohama.

nese, using the plebeian classes for rank and file, have carried the two services to a state of the highest organization, and have proved that they can assimilate not merely the forms but also the spirit of foreign systems. On the other hand, a visit to their factories shows machinery treated carelessly, employees so numerous that they impede rather than expedite business, and

a general lack of the precision, regularity and earnestness that characterize successful industrial enterprises in Europe and America. Achievement in one direction and comparative failure in another, whereas the factors making for success are similar in each, indicates, not inca-
pacity in the latter case, but defects of standard and experience. The vast majority of the

ROCKS AND SHORE BELOW KAMAKURA.

Japanese have no ade-
quate conception of what
is meant by a highly
organized industrial or
commercial enterprise.
They have never made
the practical acquaint-
ance of anything of the
kind, nor ever breathed
a pure business atmos-
phere. For elaborating
their military and naval
systems they had close
access to foreign models,
every detail of which
could be carefully scru-

tinized, and they availed themselves freely of the assistance of foreign experts, French,
German and British. But in the field of manufacture and trade their inspection of foreign
models is necessarily superficial, and they are without the coöperation of foreign experts.
It may be supposed that, since the foreign middleman plays such an important part in the
country's over-sea commerce, his skill and experience must have been equally available for
the purposes of industrial enterprise. But two difficulties stood in the way; one legal, the
other sentimental. The treaties forbade foreigners to hold real estate or engage in business
outside the limits of the settlements, thus rendering it impossible for them either to start
factories on their own account or to enter into partnership with native industrials; and an almost
morbid anxiety to prove their independent competence impelled the Japanese to dispense
prematurely with the services of foreign employees. Rapid as has been the country's material
progress, it might have been at once quicker and sounder had these restrictive treaties been
revised a dozen years earlier, when Japan was still upon the threshold of her manufacturing career, and before repeated failures to obtain considerate treatment at the hands of
Western Powers had prejudiced her against foreigners in all capacities. In 1885 she was
ready to welcome the Occidental to every part of the country; regarded it as a matter of
course that he should own real estate, and would gladly have become his partner in commerce
or manufacture. In 1895 she had come to suspect that closer association with him might

have dangers and disadvantages, and that the soil of Japan ought to be preserved from falling into his possession. There are evidences that this mood, so injurious to her own interests, is being replaced by more liberal sentiments; but in the meanwhile she has been induced to stand aloof from alien aids at a time when they might have profited her immensely, and to struggle without guidance toward standards of which she has as yet only a dim perception. Already, too, some of the advantages of cheap labor and inexpensive living are disappearing, and, on the whole, there seems to be little doubt that though great manufacturing successes lie before her, she will take many years to realize them.

CEREMONY ON THE ARRIVAL AND DEPARTURE OF A GUEST.

MODERN JAPAN IN BRIEF.

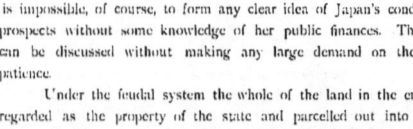

T is impossible, of course, to form any clear idea of Japan's condition and prospects without some knowledge of her public finances. The subject can be discussed without making any large demand on the reader's patience.

Under the feudal system the whole of the land in the empire was regarded as the property of the state and parcelled out into two hundred and seventy-seven fiefs, great and small, which were assigned to as many nobles. These held the land in trust and were empowered to derive revenue from it on the basis of one half of the produce to the feudal chief, one half to the farmer. The latter was only a tenant. He had no title to the soil he tilled and might not transfer it. But he generally received considerate treatment from the officers of the fief, and could count on almost absolute fixity of tenure. In practice it was found convenient to collect the revenues of the fiefs in the form of rice only, leaving the other crops entirely to the farmer. Thus the whole burden being thrown on the rice crop, the proportion of half to the feudatory and half to the farmer did not hold in the case of that cereal. Out of a total production of one hundred and fifty-three million bushels, the portion delivered for the support of the nobles and their retainers aggregated one hundred and twelve million bushels annually, the money value of which at the time of the Restoration (1867) was one hundred and sixty-five million *yen*.[1] The fiefs having been handed back to the sovereign in 1871, it was decided to provide for the feudal nobles and the *Samurai* in general by the payment of lump sums in commutation or by handing to them public bonds the interest on which should constitute a source of income. The result of this transaction, into the details of which we need not enter, was that bonds having a total face value of one hundred and ninety-one and a half million *yen* were issued, and ready-money payments aggregating twenty-one and a quarter million *yen* were made. This was the foundation of Japan's national debt. Indeed these public bonds aggregating one hundred and ninety-one and a half million *yen* may be said to represent the bulk of the state's liabilities during the first twenty-five years of the *Meiji* period. The government had also to take over the debts of the fiefs, amounting to thirty-one

[1] The *yen* of these calculations is the present unit of Japanese currency — a coin worth fifty gold cents (two shillings), approximately.

million *yen*, of which twenty-one and a half millions were paid with interest-bearing bonds, the remainder with ready money. If to the above figures we add two foreign loans aggregating sixteen and a half million *yen* (now completely repaid), a loan of fifteen million *yen* incurred on account of the only serious rebellion that marked the passage from the old to the

new *régime*,— the Satsuma revolt of 1877,— loans of thirty-three million *yen* for public works, thirteen million *yen* for naval construction and fourteen and a half millions in connection with the fiat currency, we have a total of three hundred and five million *yen*, being the whole national debt of Japan during the first twenty years of her new era under Imperial administration.

SMALL SUMMER HOTEL AT KANAZAWA.

We need not further examine the origin and early history of the country's debt. The story is sufficiently set forth in the above figures. Let us pass at once then to the great epoch of Japanese finance—the war with China in 1894-95. The direct expenditures on account of the war aggregated two hundred and forty million *yen*, of which total one hundred and thirty-five millions were added to the national debt, the remainder being defrayed with accumulations of surplus revenue, with a part of the indemnity received from China and with voluntary contributions from patriotic subjects. In the immediate sequel of the war the government elaborated a large programme of armament expansion and public works. The army at the time of the war consisted of six divisions and the Imperial guards, with a peace establishment of seventy thousand and a war strength of two hundred and sixty-eight thousand; the navy of thirty-three vessels, exclusive of twenty-six torpedo boats, representing a displacement of sixty-three thousand tons. It was resolved to raise the number of divisions to twelve, with a peace establishment of one hundred and forty-five thousand and a war strength of five hundred and sixty thousand, and the navy to sixty-seven ships (besides eleven torpedo catchers and one hundred and fifteen torpedo boats) with an aggregate displacement of two hundred and fifty-eight thousand tons. The expenditures for these unproductive purposes as well as for coast fortifications, dockyards, and so on, came to three hundred and twenty-five million *yen*, and the total of the productive expenditures included in the programme was one hundred and ninety million *yen;* namely, one hundred and twenty millions for railways, telegraphs and telephones, twenty millions for riparian improvements, twenty millions in aid of industrial and agricultural banks and so forth; the whole programme thus involving an outlay of five hundred and fifteen million *yen.* To meet this large figure the Chinese indemnity

and other assets furnished three hundred millions, and it was decided that the remaining two hundred and fifteen millions should be obtained by domestic loans, the programme to be carried completely into operation, with trifling exceptions, by the year 1905. This somewhat wearisome array of figures may be concluded by adding that Japan's total indebted-

REST-HOUSE ON NOGE HILL, ABOVE YOKOHAMA, IN CHERRY-BLOSSOM TIME.

ness is now four hundred and thirty-five million *yen* in round numbers; that it will reach its maximum—four hundred and ninety-seven millions—in 1902, and that the country has no longer any foreign debt.

Having now obtained a clear idea of the state's liabilities, we pass to consider briefly its assets. The public revenue is remarkably small when viewed with regard to the population (forty-two and a half millions) and to the career of vigorous progress upon which the country has embarked. Only one hundred and twenty-eight million *yen* come into the treasury yearly, ninety-three and a half millions of which represent direct taxes, being at the rate of two and one fifth *yen* (one dollar and ten cents) per head of the population.

The question at once arises, how can a country maintaining an army of over half a million men and a navy of sixty-seven ships, hope to pay its way with an annual income of sixty-five million dollars? The feat presents an impossible aspect to Western financiers. But we have to note that a little money goes a long way in Japan. Her army, for example, when brought up to its new standard (in 1902) will require a yearly outlay of only twenty-seven million *yen,* and the navy. No other country supports such establishments at such a petty cost. Still, if we add thirteen million *yen* on account of maintaining the strength of the navy, and if we observe that the service of the national debt—interest and amortization—will soon stand at thirty-five million *yen,* it is seen that the total outlay under these three headings, Army, Navy, and National Debt, will by and by aggregate ninety million *yen*, thus reducing to thirty-eight millions the portion of the " revenue available for administrative expenses." Now the latter at present aggregate forty-eight millions, and will certainly amount to at least sixty millions when grants in aid of Formosan development and maritime enterprise reach their full figures. From this estimate it would appear that the state will soon be confronted with a deficiency of revenue amounting to twenty-two million *yen*, if it adheres to the programme of armament expansion and productive public works mapped out in the immediate sequel of the war with China. This question, though its outlines have been but vaguely appreciated, has of late caused considerable uneasiness in Japan and evoked pessimistic criticisms abroad. A very brief examination will show that no grounds exist for uneasiness. Omitting minor items of expanding income—such as a steady growth of the receipts from state railways, posts and telegraphs; an addition of fully seven million *yen* from customs duties under the new tariff, and so forth—there is the general fact that a large reserve of taxable capacity exists. It has been shown above that the farmers formerly paid to their immediate landlords, the feudal nobles, fully two thirds of the rice crop, which payment represented rent and taxes. When feudalism fell (in 1871) the nobles ceased to stand between the sovereign owner of the soil and the agricultural classes that tilled it. The Emperor then took a remarkable step. He conferred on the farmers, who had hitherto been mere tenants, the right of permanent possession, in consideration of a perpetual annual payment nominally representing two and a half per cent of the market value of the land, but in reality amounting to less than one and a quarter per cent. The result was that eleven million acres of land became the property of the farmers with-

FUJIYAMA FROM KAMADO NEAR GOTENBA, LATE IN THE FALL.

out the disbursement of any purchase money on their part, their sole liability being a yearly payment of three and a half *yen* (one dollar and seventy-five cents) per acre. It is to this payment that the name of " land tax " is now given; a misleading term, since it conceals the fact that rent also is included in the sum. The average rice crop at present is two hundred million bushels, and if two thirds of it were paid into the treasury, as was the case in feudal times, the resulting revenue derived by the state from the land would be about two hundred and sixty-five million *yen*, whereas it is only thirty-eight and a half millions under the present system. In short, the landholder pays about one seventh of what he paid formerly, and he owns the land into the bargain. It does not appear that he fared very badly under the old system, and certainly, judging from past experience, there can be no doubt of his ability to contribute much more liberally to the public revenue than he does at present. But there is great reluctance to make any additional demand on the agriculturist. Taxation in former times signified solely the payments made by farmers. Mechanics and tradesmen were not regularly taxed. The system would have been exactly that of the "single tax" had not the urban lands escaped, as they do still escape, with a most unjustly small impost. Hence to reduce taxation meant, in practice, to abate the agriculturists' contributions; and inasmuch as the Emperor promised to lighten the people's burdens after the abolition of feudalism, his pledge was held to have been addressed direct to landholders. Thus a sentimental notion exists that immunity has been imperially guaranteed to tillers of the soil, and no cabinet nor any political party cares to imperil its popularity by declaring for an increase of the land tax. The facts, however, cannot be gainsaid, and by and by the present romantic mood will yield to plain expediency. Then there are the taxes upon alcoholic beverages and upon tobacco. By taxing the former at the rate of twenty-five cents a gallon and subjecting the latter to an impost of ten cents a pound, the government could obtain revenues of seventy-two million *yen* and eighteen millions where it now obtains only thirty millions and eight millions, respectively. In a word, there would be no difficulty whatever in raising from direct taxes an income of one hundred and eighty million *yen* annually, instead of ninety-four millions now raised. It is a great misconception to speak of the Japanese as heavily taxed. The fact is that for the past twenty-seven years they have been accustomed to taxes so light that they cannot easily reconcile themselves now to the prospect of a less agreeable state of affairs. Still timid in their ideas of finance, they have not fully realized that their new status imposes a greatly altered scale of national expenditure. But the ability to contribute much more liberally to the funds of the state than they do at present certainly exists, and can be invoked at any time.

The great difficulty under which Japan labors is want of capital. Statistics show that the capital actually engaged in public and private enterprises is five hundred and ninety million *yen*, and that seven hundred and eighty-four millions are pledged, though not yet paid up. On the other hand, the volume of circulating media is only three hundred million *yen*,[1] and

---

[1] That is to say, between seven and eight *yen* per head of the population. It is interesting to note the corresponding figures in Occidental countries; thus: Austria, nineteen *yen*; Denmark, twenty-five; France, sixty-five; Germany, thirty-three; England, thirty-three; the United States, forty-six. There has been much talk on the part of superficial Japanese and foreign critics about an excessive volume of currency in Japan, but it will be seen from the above figures that the contention is quite baseless.

THEATRICAL PERFORMANCE.

Scene from an historical play dealing with ancient times entitled "Gosho Guruma," literally "The Imperial Cart." The figure in white standing upon the cart represents the Mikado of the time. The Japanese theatre is the only place where the life of Old Japan can be studied in these radical latter days. The acting, costumes and scenic effects are utterly unlike anything seen outside the Orient.

the funds at the disposal of the banks total three hundred and twenty-four millions. Recourse to foreign capital would be the natural plan under such circumstances, but, as already explained, so long as she was a silver-using country, Japan hesitated to contract gold debts, and the capitalists of Europe or America would of course have insisted that all loans should be on a gold basis. It was chiefly with the object of removing this obstacle that Japanese financiers decided, in 1897, to adopt the gold standard. They hoped that their five-per-cent public bonds would thus become readily salable in the markets of Europe and America. But Europe and America have not yet acquired full confidence in Japanese finance, and, moreover, the bonds themselves present some features not unlikely to deter foreign investors. The day is probably not far distant when Japanese securities will become favorite investments abroad. Meanwhile there is certainly a great opening for foreign capital in the field of industrial enterprise. An examination of the returns of sixty-eight joint-stock companies for the second half of 1897 shows that they paid an average dividend of sixteen and an eighth per cent, and it is not to be doubted that still better results could be attained were foreign business experience and cheap capital available.

One of the paradoxes of modern Japan's story is that her political advance has proved a barrier to her material progress. In the great national fermentation that inaugurated the *Meiji* era, the state's best men rose to the surface, and had it been possible to leave the reins of government permanently in their hands, a steady march along all lines of sound development would have been assured. But it was not possible. Parliamentary government had to come, bringing with it all the confusion of multitudinous counsels and the clashing of party politics. Let us trace the record briefly.

PLAYING BLINDMAN'S BUFF IN A SIDE STREET LEADING TO THE MOAT IN TOKYO.

No one reading Japanese history carefully can fail to infer that representative institutions are in the genius of the nation. From the very earliest eras the sovereign ceased to be autocratic. All the highest offices of state became the hereditary possessions of certain great families, and as generation followed generation each unit of this oligarchy of households attained the dimensions of a clan. By and by the exigencies of the times gave birth to a military aristocracy, headed by a generalissimo (*Shogun*), into whose hands the administrative authority passed. But even in this military feudalism no traces of genuine autocracy are found. Just as the extensive powers nominally vested in the central figure, the *Shogun*, were

in reality wielded by a large body of ministers and councillors, so the local autonomy enjoyed by each fief was exercised, not by the chief himself, but by his leading vassals. A united effort on the part of all the clans to overthrow such a system and wrest the administrative power from the *Shogun* could have only one logical outcome, the combined exercise of the

BRONZE HORSE AT SUWA TEMPLE, NAGASAKI.

recovered power by those who had been instrumental in recovering it. There was no open enunciation of such a principle, but it could be read clearly enough on the face of events and between the lines of the oath sworn by the Emperor when he placed himself once again at the head of civil affairs in 1867 — the oath that " wide counsels should be sought and all things determined by public discussion." It is plain that to fulfil this promise a representative assembly must be convened. But when? Some preparations were indispensable, and after a time the nation began to suspect that the preparatory interval was being unduly prolonged in the interests of the clique of statesmen whom the Restoration had brought into office, the leading clansmen of Satsuma and Choshiu. At first the suspicion found only vague expression, but by degrees it grew into a dread, partly genuine, partly factitious, that two or three great clans were seeking to climb into permanent power on the ruins of feudalism. A clamor against *Sat-Cho* (Satsuma and Choshiu) cabinets then (1878) began to be audible, and found a mouthpiece in a now celebrated politician, Count Itagaki, who had seceded three years previously from the ranks of the Restoration statesmen for reasons that need not concern the reader. At the time of his secession his love of liberty had not hardened into a political platform. That came a little later, and with it came occupants of the platform — a few who had been in office and wanted to repeat the experience, a greater number who had never been in office and wanted to try the experience, and a still greater number who wanted freedom of speech and representative institutions. Thus the Liberal party (*Jiyu-to*) was organized, the first political association of modern Japan. Three years later (1881) another notable secession from the ranks of officialdom took place. The seceder was Count Okuma. He, too, wanted representative institutions, but he wanted them at once. That was the chief difference between him and Count Itagaki. The latter labored for the principle, the former for its immediate practical application. It may be supposed that the two would thenceforth have joined hands, and indeed, so far as theory was concerned, nothing held them apart. But officially they were enemies. Itagaki seceded in 1875 and formed his party in 1878. Okuma seceded in 1881 and formed his party at once. Thus Okuma had remained for six years a member of the cabinet that threw off Itagaki, and in his capacity of minister had during that time been privy to sundry severe measures for the restraint of Itagaki's agitation. So the two men, though working for the same purpose with similar instruments, could not join forces. Okuma called his followers the Progressists, a term well adapted to the circumstances of the party's birth.

These names deserve notice. They afford a clew to one of the chief obstacles standing in the way of the consummation of representative institutions – party government. It is plain that a true Progressist must be a Liberal, and a true Liberal a Progressist. The terms are distinctive only; they do not indicate any difference of platforms. There is, in fact, no difference and there can be no difference. The Japan of to-day is permeated from head to foot by the spirit of progress. If a rare conservative survive here and there, he has not strength to make his existence practically noticeable. Parties are grouped about persons, not about principles. It is thus inevitable that their elements should lack cohesion and be always subject to disintegrating influences. Another difficulty, more remediable, but for the moment not less serious, is that no party has yet succeeded in permanently attaching to itself any considerable number of the statesmen whose experience indicates them as the natural and indeed the only trustworthy managers of state affairs. That is a point demanding brief explanation, for to Anglo-Saxon readers it will appear scarcely comprehensible that statesmen and politicians should be grouped in different camps. The records of England's pre-Victorian struggle for party government shows us a king on one side, clinging to his traditional prerogative; a parliament and its leaders on the other, fighting against subservience to the royal mandate. But in Japan the contest has always been between officialdom and non-officialdom. At the Restoration in 1867, as has been already noted, a group of the most brilliant and competent men in the Empire got the administrative reins into their hands and held them for twenty-four years. At first they were virtually unopposed. A few adversaries might have been found among lovers of the old order; but these had no strength to raise their heads, and were, moreover, hopelessly out of touch with the time. By degrees, however, the vast hotchpotch of changes and reforms introduced by the hands of eager statesmen became a dish too varied to suit the palate of all. Now by this measure, now by that, tastes were differentiated, and unit after unit of the administrative coterie drifted into opposition. A few had recourse to force, and fell in battle or under the headsman's sword. But, with these few exceptions, peaceable opposition was the programme of the seceders. They were, in fact, advocates of constitutional government, and they endeavored to limit their advocacy to constitutional methods. Foremost among them were the two spoken of above, Counts Itagaki and Okuma. These men alone need be considered in connection with the growth of political parties; their figures dwarf all the rest. Count Itagaki is the Rousseau of Japan. A passionate lover of liberty in the abstract, he does not concern himself deeply about the concrete applications of his principles. Count Okuma may be compared in some respects to Sir Robert Peel. To remarkable financial ability and a lucid, vigorous judgment he adds the faculty of placing himself on the crest of any wave that a genuine *aura popularis* has begun to swell. Japanese political parties are often spoken of by foreign observers as something quite beyond ordinary comprehension. But the story of their evolution presents no real difficulty. The two here mentioned— Liberals and Progressists —are the only associations of the kind that need be seriously considered by the student. Cabals have been formed within their ranks and independent coteries outside, the occasional result being a welter of factions, bewildering and disheartening. But such things represent the incidental struggles of the moment, not the guiding principles of the era. Officialdom stood arrayed against the parties; officialdom under the leadership of the Restoration statesmen, with whom were allied the army officered chiefly by Choshiu men, and the navy officered chiefly by Satsuma men. On one side, two strong bodies of political agitators struggling to obtain the practical fulfilment of the Emperor's promise; on the other, the Sat-Cho holders of office and all their official followers, struggling to postpone that consummation, —such was the spectacle presented to the nation. It might easily have become a dangerous spectacle, but the government paralyzed its elements of commotion by proclaiming (in 1881) that a constitution should be issued in 1890 and a Diet convened in 1891. Thenceforth the parties could only wait. The framer of the constitution was Marquis Ito. These titles, "Marquis," "Count," "Viscount," must not be taken as indicating that the makers of Japan's modern history belonged to the ancient nobility of Japan. Marquis Ito was plain Mr. Ito when he first appeared upon the political stage, and the same is true of Okuma, Inouye, Itagaki, Kido, Saigo, and so on; in short, of nearly the whole group of brilliant publicists who led Japan from the old to the new and steered her through all her subsequent difficulties. Many of them can look back to strange and stirring experiences since the days when, as feudal retainers, with no foundations for fortune-building except high courage and keen intelligence, they laid plans that must have seemed at the moment idle dreams, but were destined to raise the country to an unlooked-for place among the nations, and themselves to heights of influence and fame such as their most ambitious fancies cannot have pictured. Count Inouye, who holds the portfolio of finance at the time when these pages are written, remembers how, one winter's night thirty-five years ago, he and a little band of "patriots" applied the torch to the new buildings destined for the British Legation in Tokyo and burned them to the ground. Marquis Ito can recall how he, with the same Count Inouye, made the journey to England as sailors before the mast in 1862, and arrived in London with barely sufficient money to buy a loaf of bread. Marquis Saigo has not forgotten the fight he and his comrades waged thirty-six years ago in the upper story of an inn, when they had to choose between death and abandonment of their anti-foreign crusade. Baron Morioka, who died recently after having occupied many important public posts, was one of two *samurai* who slashed at three foreigners on the Tokaido in 1862, killing one and wounding the others, which event led to the bombardment of Kagoshima by a British squadron the following year.

Such records are numberless. Nothing is stranger in the story of new Japan than to compare the period of "blood and iron" when these warrior statesmen made their *début* with the era of peaceful progress they have introduced. Marquis Ito will probably leave the most enduring fame. He is at once the statesman and the legislator of his time. To him the country owes its first constitution, promulgated in 1890 and proudly pointed to by the Japanese nation as the only charter of the kind voluntarily given by a sovereign to his subjects. In other countries such concessions have always been the outcome of long struggles between ruler and ruled; in Japan the Emperor freely divested himself of a portion of his prerogatives and transferred them to the people. That view of the case is, of course, not untinged with romance, as will be seen from what has been written above about the growth of political parties, but on the whole its truth cannot be questioned. Marquis Ito and the jurisconsults who helped him to frame the constitution did not err on the side of rashness. They

THE GRAVES OF THE FORTY-SEVEN RONINS.

The tombstone of the leader (Oishi Kuranosuke) is in the rear corner. Evergreen sprigs are kept in the bamboo receptacles the entire year by the people, not by any organization.

fixed the minimum age for franchise-holders and parliamentary candidates at twenty-five and the property qualification for each alike at payment of fifteen *yen* annually in the form of direct taxes, which meant an income of some twelve hundred *yen*. The result of the tax-paying limit was that only four hundred and sixty thousand persons were qualified to elect and be elected out of a population numbering sixteen million males of the required age. A bicameral system was adopted for the national assembly: the House of Representatives, numbering three hundred members; the House of Peers, partly hereditary, partly elective and partly nominated by the sovereign. No incident in Japan's modern career was watched with more curiosity than this sudden plunge into parliamentary institutions. There had, indeed, been some preparation. Provincial assemblies—which, as part of the systems of local autonomy gradually introduced after the fall of feudalism, were themselves an innovation—had familiarized the people more or less with the methods of legislative bodies. But provincial assemblies were at best petty arenas—places where the making or mending of roads and the policing and scavenging of villages came up for discussion, and where political parties found no opportunity to attack the government or to debate problems of national interest. Besides, not much was known about provincial assemblies. Only the briefest outlines of their proceedings appeared in print; their legislative doings had no general interest, and a vague impression prevailed that if they were not noticed it was because they did not deserve attention. Thus the convening of a Diet and the sudden transfer of financial and legislative authority from the throne and an oligarchy of tried statesmen grouped around it to the hands of men whose qualifications for public life rested on the verdict of electors themselves apparently devoid of all light to guide their choice—this sweeping innovation seemed likely to tax severely, if not to overtax completely, the progressive capacities of the nation. What enhanced the interest of the situation was that the oligarchs who held the administrative power had taken no pains to win a following in the political field. Knowing that the opening of the Diet would be a veritable letting loose of the dogs of war, an unmuzzling of the agitators whose mouths had hitherto been closed by legal restrictions upon free speech, but who would now enjoy complete immunity, whatever the nature of their utterances, within the walls of the assembly—foreseeing all this, the statesmen of the day nevertheless stood severely aloof from alliances of all kinds, and discharged their administrative functions with apparent indifference to the changes that popular representation could not fail to induce. That somewhat inexplicable display of unconcern became partially intelligible when the constitution was promulgated, for it then appeared that the cabinet's tenure of office was to depend solely on the Emperor's will; that ministers were to take their mandate from the throne, not from parliament. Here was a fresh illustration of the theory underlying every part of the Japanese polity. Laws might be redrafted, institutions remodelled, systems recast, but amid all changes and mutations one steady point must be carefully preserved, the throne. The makers of new Japan understood that so long as the sanctity and inviolability of the Imperial prerogatives could be preserved, the nation would be held by a strong anchor from drifting into dangerous waters. They labored under no misapprehension about the inevitable issue of their work in framing the constitution. They knew very well that party cabinets are an essential outcome of representative institutions, and that to party cabinets Japan must come. But they regarded the

Imperial mandate as a conservative safeguard pending the organization and education of parties competent to form cabinets. Such parties did not yet exist, and, until they came into unequivocal existence, the Restoration statesmen, who had so successfully managed the affairs of the nation during a quarter of a century, resolved that the steady point furnished by the throne must not be abandoned.

On the other hand the agitators found here a new platform. They had obtained a constitution and a Diet, but they had not obtained an instrument for pulling down the "clan" administrators, since these stood secure from attack under the ægis of the sovereign's mandate. They dared not raise their voices against the

STONE ARCH ON MITAKE YAMA, USUALLY CALLED ONTAKE.
This is one of the 'oldest mountains in Japan, and, next to Fuji, the most sacred.

unfettered exercise of the Mikado's prerogative. The nation, loyal to the core, would not have suffered such procedure, nor could the agitators themselves have found heart to adopt it. But they could read their own interpretation into the text of the constitution, and they could demonstrate practically that a cabinet not acknowledging responsibility to the legislature was virtually impotent for lawmaking purposes.

These are the broad outlines of the contest that began in the first session of the Diet and has continued ever since. The struggle presented varying aspects at different times, but the fundamental question at issue has never changed. Obstruction was the weapon of the political parties. They sought to render legislation and finance impossible for any ministry that refused to take its mandate from the majority in the Lower House, and they imparted an air of respectability and even patriotism to their destructive campaign by making "anti-clannism" their war cry, and industriously fostering the idea that the struggle lay between administration guided by public opinion and administration controlled by a clique of clansmen --Satsuma and Choshiu --who stood between the throne and the nation. There could be no doubt about the ultimate success of such tactics. At first the government showed a very resolute front. For five years it ignored the hostility of the Lower House and held by the constitutional principle of responsibility to the Emperor only. In vain the opposition threw out the budget, passed votes of want of confidence, or submitted to the throne addresses impeaching the ministry. The cabinet remained looking calmly down from its high place on the sea of tumult raging below.

It must be confessed that there was something at once sad and impressive in the spectacle. The Restoration statesmen were the men who had made modern Japan; the

men who had raised her, in the face of immense obstacles, from the position of an insignificant Oriental state to that of a formidable unit in the comity of nations; the men, finally, who had given to her a constitution and representative institutions. Yet these same men were now fiercely attacked by the arms that they had themselves nerved; were held up to public obloquy as

KEAK OF AN ORDINARY DWELLING.

self seeking usurpers, and were declared to be impeding the people's constitutional route to administrative privileges, when in reality they were only holding the breach until the people should be able to march into the citadel with some show of orderly and competent organization. That there was no corruption, no abuse of position, is not to be pretended; but, on the whole, the conservatism of the clan statesmen had only one object, to provide that the newly constructed representative machine should not be set working until its parts were duly adjusted and brought into proper gear. There is no doubt that the leaders on both sides understood the situation accurately. The heads of the parties, while they publicly clamored for parliamentary cabinets, privately confessed that they were not yet prepared to assume administrative responsibilities; and the so-called "clan statesmen," while they refused before the world to accept the Diet's mandates, admitted within official circles that the question was one of time only. It is well to note this mutual understanding, for its existence indicates that the contest must be peaceful throughout.

Little by little, too, the so-called "clan statesmen" are stepping out of the shadow of the throne and associating themselves with the political parties. When that process has been carried somewhat further, party cabinets, with all their drawbacks and disadvantages, will become an accomplished fact, and Japan will have worked her way tranquilly to a goal which other nations reached through scenes of turmoil and even violence.

The expression "tranquilly" is strictly justified by facts. It would have been reasonable to expect that tumult and intemperance must disfigure the proceedings of a Diet whose members were entirely without parliamentary experience, but not without grievances to ventilate, wrongs, real or fancied, to avenge, and abuses to redress. On the whole, however, there has been a remarkable absence of anything like disgraceful license. The politeness, the good temper and the sense of dignity which characterize the Japanese have always saved the situation when it threatened to degenerate into a "scene." Foreigners entering the House of Representatives in Tokyo for the first time might easily misinterpret some of its habits. A number distinguishes each member. It is painted in white on a wooden indicator, the latter being fastened by a hinge to the face of the member's desk. When present he sets the indicator standing upright, and lowers it when leaving the House. Permission to speak is not obtained by catching the president's eye, but by calling out the aspirant's number, and as members often emphasize their calls by hammering their desks with the indicators, there are moments of decided din. But for the rest, orderliness and decorum habitually prevail. Speeches have to be made from a rostrum. There are few displays of oratory or eloquence. The Japanese formulates his views with remarkable facility. He is absolutely free from gaucherie or self-consciousness. He can think on his feet. But his mind has never busied itself much with abstract ideas and subtleties of philosophical or religious thought. Flights of fancy, impassioned bursts of sentiment, appeals to the heart rather than to the reason of an audience, are devices strange to his moral habit. He can be rhetorical, but not eloquent. In all the parliamentary speeches hitherto achieved it would be difficult to find a passage deserving the latter epithet. From the very outset the debates have been recorded verbatim.

ON THE BEACH AT ENOSHIMA, AT LOW TIDE.

Without any forethought of parliamentary reporting, years, indeed, before the date fixed for the promulgation of the constitution, a little band of students elaborated a system of stenography, based on English models, and adapted it to the syllabary of the Japanese language. Their labors remained almost entirely without recognition or remuneration until the Diet met, when it was happily and unexpectedly discovered that a thoroughly competent staff of shorthand reporters could be organized at an hour's notice. Japan can thus boast that, alone among the countries of the world, she possesses an exact record of the proceedings of her Diet from the moment when the first word was spoken within its wall.

We may here remark that the influence of parliamentary debate upon the growth and structure of the Japanese language has been second only to the influence of the newspaper. Chiefly upon the editor and the politician has devolved the duty of presenting to the people in intelligible form the encyclopædia of new conceptions that came in the train of Western civilization. It is difficult to picture to one's self the dimensions of the task, or to believe that any language could be sufficiently elastic to furnish equivalents for such an endless terminology of absolutely novel philosophies, sciences and systems as the Occident had to offer. But the Japanese language, or, to be more accurate, the Chinese — for the reader will understand that although the pronunciation of words and the construction of sentences differ in Japan and China, the whole vocabulary of the latter country is used in the former — the Japanese language possesses extraordinary potentialities. The easiest way for an Anglo-Saxon to grasp the facts of the case is to suppose that every syllable — not every word but every syllable — of the English language had a distinct meaning of its own, and that the syllables were capable of being combined in groups of two, three, four or five to form new words. It is evident that words thus constructed might be made to express the finest shades of meaning, and that their number would be practically limitless. The Chinese ideographs are precisely such syllables. There are some twenty thousand of them immediately accessible, and by welding them together in groups of few or many, modern Japanese scholars have filled a lexicon with words which had no existence fifty years ago and would have possessed no meaning for the men of that time, but which accurately convey the sense of the foreign terms they represent. Scarcely a day passes without some addition being made to this lexicon, and the middle and upper ranks of society are becoming more and more permeated with men whose carefully constructed phrases and classical terminology smack of the journalistic article or the parliamentary debate. It is said that when the members of the first Diet read the verbatim report of their first day's proceedings in the *Official Gazette* on the following morning, they were horrified to find all their provincialisms and faults of diction mercilessly reproduced. The quality of their speech has greatly changed since then. Whether the language will undergo that much-discussed radical alteration from an ideographic to an alphabetical script is a question which some hesitate to answer and others answer in the negative, but the writer of these pages entertains no doubt whatever on the subject. Nothing marks the stages of a nation's civilization more emphatically than the nature of the vehicle it employs for transmitting ideas. The stenographist is as far removed from the hieroglyphist as the president of an American college from a Polynesian chief. It is impossible to believe that a people so essentially progressive as the Japanese will

MANGWANJI TEMPLE AT NIKKO.

One of the smaller of the twenty-eight temples at Nikko. The Japanese Buddhist temple is a modification of a Chinese modification of the Indian original. Shinto temples, like Imperial palaces and ordinary dwellings, are developments of the simple hut of pre-historic times.

permanently condemn themselves to the use of script which renders their literature a sealed book to the whole world of the West and doubles the educational difficulties that their chil dren have to overcome. From his tender years the mind of a Japanese youth is reduced to a mere memorizing machine before he has succeeded in engraving upon its tablets the seven or eight thousand ideographs which constitute the equipment of an educated man. If that great, morally injurious and comparatively useless task did not, as it certainly does, handicap him fatally in the race for general knowledge, there would still be the other objection that a nation whose written language is mechanically unintelligible to foreign peoples must always remain isolated. The anti-foreign edicts of the early Tokugawa *shoguns* have been torn up; the "barbarian-expelling" fanaticism of mediæval times has been replaced by a frank though self-asserting liberalism; the essentials of Occidental civilization have entered in to dwell side by side with the refinements and artistic etiquette of old Japan; but all the thoughts of the people in this era of wonderful progress, all the impressions that they receive from contact with the systems and sciences of Europe and America, are either completely hidden from the eyes of the nations whose intimacy they court, or find halting expression in the clumsy renderings of incompetent translators. Can such a state of affairs be permanent? The truth is that the Japanese have not yet awakened fully to the importance of this prob lem. The sense of proportion, which owes so much of its development to opportunities for observation, is necessarily defective among them. The conventionalisms of classic Chinese they understand and appreciate, but when they enter the field of general diction, they seem unable to estimate how much the force of an idea owes to the form of its expression. An industrious purveyor sits proudly under a signboard informing the public that he sells "extract of fowl" (*i. e.*, eggs); a haberdasher with a large foreign *clientèle* sports the legend " Ladies furnished in the upper story;" a blacksmith blows his bellows behind an announce ment that he has been "imstracted by French horse-leach;" the editors of a magazine for teaching English make a lady "puff at her tiny pipe in order to dissemble herself," or a startled father "roar out, taking it for a mischief of some naughty boy;" and the educational authorities go to the highways and byways to pick up "professors" of European tongues. Thus the nation is content to imagine that its mental processes and the trend of its intelli gence are adequately depicted through the agency of a host of hybrid publications the out growth of recent enterprise, newspapers, brochures and magazines, their pages inscribed with ideas which, if they do not emanate from the vapory and immature fancies of callow students, are robbed of all dignity and grace by the grotesqueness of the language employed to express them. Just as the nation fails to detect how greatly its intelligence is misrepre sented by such media, so its perception has not been aroused to the barbarous clumsiness of the ideographs and to the impenetrable barrier they oppose to free interchange of thought. But that an effective sense of these disadvantages will be born sooner or later, no observer of Japan's progress can doubt.

Materials for constructing a language adapted to parliamentary requirements, or, indeed, to the requirements of any science or philosophy, are thus abundant in Japan, but the method of procedure in the Diet is not calculated to encourage oratorical displays. Every measure of importance has to be submitted to a committee, and not until the latter's report has been received does serious debate take place. But in ninety-nine cases out of every hundred the committee's report determines the attitude of the house, and speeches are felt to be more or less superfluous. One result of this system is that business is done with a degree of celerity scarcely known in Occidental legislatures. For example, the meetings of the House of Representatives during the session 1896-97 were thirty-two, and the number of hours occupied by the sittings aggregated a hundred and sixteen. Yet the result was fifty-five bills debated and passed, several of them measures of prime importance, as the gold-standard bill, the budget, a statutory tariff bill, and so on. Such a record seems difficult to reconcile with any idea of careful legislation; but it must be remembered that although actual sittings of the Houses are comparatively few and brief, the committees remain constantly at work from morning to evening throughout the nine weeks of the session's duration. Another interesting feature of the system is that the members of the government do not lead their supporters in the Diet or break lances with their opponents. The ministers of state and the delegates from each administrative department are entitled to be present and to address the Houses whenever they please. They avail themselves of the privilege to the extent of explaining official bills or answering questions. But they never take any part in a debate or make controversial speeches. It is a somewhat lame system; for while it brings the members of the cabinet and the government's delegates within range of the opposition's invective, it does not enable them to exert any sensible influence on a debate. Nevertheless there is no gainsaying the fact that the legislative laurels have been entirely by the government, session after session. Thus, in the 1896-97 session alluded to above, out of fifty-five measures debated and passed only three were private bills — a strong corroboration of the

TYPE OF THE GEISHA CLASS.

criticism that whatever destructive capacity has hitherto been displayed by Japanese political parties, their competence for purposes of constructive statesmanship remains an entirely undemonstrated quality.

That the aspiration common to political parties the world over, namely, parliamentary cabinets, would find expression in Japan also, might have been foretold by any student of constitutional history. But we have to answer the more complicated and interesting question, what is the ultimate tendency of political thought in Japan, and what will be the final lines of party cleavage? It must be confessed at once that these points are still wrapped in much obscurity. With the records of eleven sessions of the Diet and the platform utterances of many years to guide us, there should not be any insuperable difficulty in drawing definite inferences. Yet, when we proceed to catalogue the objects for which politicians have hitherto fought, we find that the list does not include anything particularly suggestive. Reduction of the land tax, freedom of speech and pen, party cabinets, petty modifications of the local-government system, abuses of official power, acts of financial dishonesty, administrative reform, treaty revision and foreign affairs,--such are the matters that have chiefly occupied political attention. In view of what has already been written about taxation in general and the land tax in particular, it may seem strange that a reduction of the agricultural classes' burdens should have been gravely advocated by any association of politicians. But in the early days of Japan's parliamentary career the people's representatives, posing for the first time in that character, were naturally anxious to play the part of popular benefactors. Lighter taxes were the most obvious means to that end, and as from time immemorial the nation's contributions to the administration had been virtually limited to the land tax, the latter suggested itself as the proper object of attack, quite apart from fiscal justice or expediency. An incident so commonplace would scarcely be worth special mention had it not led to curious complications. In order to cut down the land tax some other source of revenue had to be found, or some retrenchment of expenditures had to be effected. The latter method was chosen, and, as a matter of course, the reformers concluded that the proper domain to be invaded was that of official salaries. Now the number of officials engaged in administering Japan's affairs is forty-two thousand seven hundred and twenty-eight, and their yearly stipends total thirteen million six hundred and sixty-eight thousand two hundred and forty-six *yen*, so that their average monthly pay is less than eleven dollars (gold). There does not seem to be much room for retrenchment there, and the case presents itself in a still more striking light when we note that the one thousand six hundred and seventy-four judges and public procurators who constitute the judiciary receive an average salary of thirty-six and one half dollars (gold) a month, and that the seventy-three thousand one hundred and sixty teachers engaged in the primary schools have to make ends meet with a monthly pittance of a little over three and one half dollars. It is plain that official integrity and competence cannot possibly be looked for under such circumstances. When a common factory hand can earn more than a school teacher, and when a clerk in a store gets as much as a judge on the bench, the judicial and educational careers will never attract men of talent. In fact, Japan's present attitude is one of financial shrinking from the responsibilities of her new career. She is like a youth who engages in every kind of exercise calculated to promote his growth, yet seeks all the while to clothe himself in the garments of his boyhood. Her administrators are not free from peculation, her judiciary from corruption, her officials from incompetence; nor will they ever be free until the emoluments of office become more important than its opportunities. Blind to that rudimentary truth, the opposition in the first House of Representatives—and practically the whole House was in opposition—made a fierce onset upon official salaries. They had no right to meddle with the matter: the constitution explicitly reserves to the sovereign the prerogative of appointing and dismissing officials and fixing their stipends. But the House, as we have said, was in a destructive mood. If it might not constitutionally legislate for a reduction of salaries, it could at any rate refuse to pass any budget including payments on the old scale. That was what it did. Probably these party politicians had no immediate object beyond providing means to reduce the land tax. But they incidentally achieved another result: they placed the so-called "clan statesmen" in the doubly invidious position of clinging to power in the interest of the clans as opposed to the interest of the nation at large, and of refusing to abate anything of their own emoluments for the sake of lightening the burdens of the

VIEW OF KATSURA RIVER NEAR ARASHIYAMA.
The path leads to Kyoto. The logs have been floated down from the province of Tamba.

people. It says much for the resolution and foresight of those statesmen that they stood firm under such circumstances. The event has justified them, for to-day the nation recognizes that a programme precisely the converse of that advocated by the people's representatives seven years ago must be adopted, the land tax must be increased and official salaries must be augmented. This chapter of parliamentary history indicates that political wisdom is still monopolized by the men whom the tumult of the revolution brought to the surface thirty years ago, and who have remained on the surface ever since. With them victory has rested from point to point of the struggle against the party politicians. Administrative reform, for example, has always been a principal plank in the platform of every political party, but no party has ever yet clearly explained its conception of reform, and the *Meiji* statesmen, by quietly waiting for an explanation, have seen the agitation perish from anæmia of ideas. Similarly in the case of treaty revision the party leaders would have pushed the government out of the broad path of patient and conciliatory effort into a narrow groove of retaliation and reprisal. But the government never wavered, and the goal was ultimately reached without any sacrifice of national dignity or loss of foreign friendship. In short, Japan's best hope of attaining the international position whither her gaze is directed seems to centre upon the band of statesmen who still enjoy the confidence of the sovereign, and upon the younger officials who, educated under their instruction, are now gradually stepping into their places. The reader will easily understand, therefore, how essential it is to the political parties that they should succeed in enrolling these *Meiji* statesmen in their ranks. There have been indications that such success was on the point of achievement. In 1895, Marquis Ito, then Prime Minister, stepped out of the shadow of the throne and formed a coalition with the Liberal party. The proximate purpose of the union was to secure the Diet's assent to a large programme of measures— including naval and military expansion — mapped out by the cabinet in the sequel of the war with China. But it is certain that Marquis Ito, the framer of the constitution, the leading statesman of Japan, the Emperor's most trusted minister, did not for a moment mistake the significance of his own act. He knew exactly what was involved in an open alliance between the cabinet and a political party; knew that it must be interpreted by the nation as a confession of the cabinet's inability to conduct state affairs without the aid of a majority in the Lower House; knew that it indicated an advance almost to the very threshold of the English system of parliamentary mandate and imperial indorsement; and knew that his association with the Liberals supplied the lack which had hitherto incapacitated them as serious candidates for office, the lack of administrative experience and prestige. It seemed now that the struggle had virtually come to an end, and that the principle of party cabinets had received practical recognition. But less than three years afterwards the same Marquis Ito found himself at the head of a cabinet with the same Liberals in opposition. His predecessor in the premiership, Count Matsukata, had assumed office with the support of the Progressists, and had resigned it with the same Progressists in opposition.

What is the explanation of these things? Simply this, that on the one hand the political parties have been so long trained in habits of irresponsible criticism as to be temporarily unfit for the duties of responsible alliance, and that, on the other, since their allegiance is to persons, not principles, they are without any strong force of cohesion or working sense of discipline. By what processes of education they are to be fitted for the *rôle* they aspire to fill, it is impossible to predict, but in the working out of every Japanese problem allowance must always be made for those factors of versatility and adaptiveness which have hitherto helped the nation so signally through all its difficulties. Some students of the time predict that the issue will be a falling asunder of the two great clans, Satsuma and Choshiu, and that their rupture will mark the line of division in the new political field; the Satsuma men, on one side, standing independent of political parties and ruling it may be by military force; the Choshiu men ranged on the other, leading the political parties and fighting for parliamentary cabinets.   To the writer such a forecast seems superficial.   The Satsuma men are richly endowed with the opportunist faculty.   Their position in this final act of the Restoration drama finds an exact analogy in the part they played at the rising of the curtain thirty-five years ago.   When the national mind was beginning to seethe with the idea of abolishing military feudalism and restoring the administrative power to the sovereign as a preliminary step to representative institutions, and when the Choshiu men stood forth as champions of the great change, the Satsuma clan joined hands with the Aizu to crush the movement and to drive its supporters from the capital; but when the tendency of the era could no longer be mistaken, Satsuma turned and combined with Choshiu to annihilate Aizu.   Proceedings of that kind were perfectly consistent with the historical character of the southern clan.   Its leaders are not quick to read the signs of the times, but they are conspicuously quick to obey the legend when the writing has become clear.   If at one moment (1897) in the *Meiji* era circumstances forced them away from their old allies and, by separating them from the two great political parties, placed them in the false position of opposing parliamentary cabinets, there was no difficulty in foreseeing that the expediency of advocating parliamentary cabinets would presently draw them back to the alliance, as, indeed, the event proved.   Is it, then, in the field of foreign politics that the dividing lines of public opinion are likely to be finally traced?   At first sight an affirmative answer suggests itself, for during the past six years a very distinct differentiation between " stalwarts " and " moderates " has been apparent.   But here, too, a bewildering feature disturbs our calculations.   The so-called " stalwarts " show a marked disposition to condemn the extensive programme of armament expansion which the " moderates " uncompromisingly support.   So, in fine, we rise from every examination of Japanese politics with a conviction that if the tests employed for analytical purposes in Western countries are applicable here, the non-official section of the people has not learned how to adapt its methods to its fancies, or acquired any sober idea of the responsibilities attaching to the control of national affairs.   After all, the solid ballast that keeps the foreign policy of every Western state on an even keel is self-interest.   Moral principles are regarded as mere deck cargo, to be thrown overboard without compunction in rough waters.   The Japanese have not risen fully to that elevated canon of international practice.   They still cherish a romantic idea that a nation, like an individual, should endeavor to tread the "*samurai's* road," the *bushi-do* of feudal days, on which the four conspicuous finger-posts are loyalty, truth, magnanimity and courage.   Though not generally credited it is nevertheless true that ninety-nine out of every hundred Japanese frankly regarded the recent war with China as a chivalrous effort on their country's part to secure the independence of little weak Korea against the present menace of big strong China's grasp and the ominous shadow of future Russian aggression. Side by side with that conviction was a passionate longing to win the credit of rousing both China and Korea from their blind conservatism, and thus obtaining for Japan world-wide recognition as the propagator of Occidental civilization in the Orient.

It may occur to the reader that the much-talked-of "atrocities" perpetrated by Japanese troops after the capture of Port Arthur fit badly into the context of a chivalrous war. That is very true. But, in the first place, the doings of the victors at Port Arthur were grossly exaggerated by newspaper correspondents with "axes of their own to grind;"

THE HOT SPRINGS AT ATAMI, ONE OF THE MOST POPULAR WATERING-PLACES.

in the second, they represented an outburst of rudimentary passion under the incentive of revolting cruelties previously committed by the Chinese; and in the third, they were the one blot on an otherwise white record of warfare conducted in strict accord with the best principles of civilization,—if such a paradox as civilized warfare be possible,—warfare absolutely free from the disgraces of private property destroyed, cities sacked or looted, women outraged or prisoners butchered. When the fighting had ended, Japan set herself with more enthusiasm than prudence to lead Korea into the path of progress, and her failure was not more bitter than the conviction subsequently forced upon her that, in exposing China's helplessness, she had virtually issued an invitation to Western powers to approach and tear her once revered neighbor to pieces. It was perfectly consistent with Japanese character, consistent with the precepts of the *bushi-do*, that the sight of her former enemy's misfortunes should inspire her with a wish to come to the helpless Empire's assistance, and it is thus that she finds herself to-day strongly drawn towards England, whose policy of preserving the integrity of China she heartily approves, and towards the United States, whose freedom from self-seeking aggressiveness she has always admired. This verdancy of international sentiment begins to disquiet Japan's political philosophers. They are proud of it, but they see its dangers, and are commencing to preach the doctrine of self-interest which the astute nations of the West accept as a guide. The national mandate to redress wrong in a superior, to defend the weak against the strong, and to sacrifice profit to honor, which made the feudal *samurai* such an interestingly chivalrous figure, which banishes the *ego* from Japanese social intercourse, which impels the student to sacrifice his scholastic career for the sake of asserting a principle, the *soshi* to imagine that a swashbuckler may derive respectability from his mission, and the patriot to regard assassination and suicide as legitimate weapons,— this natural mandate may be terribly embarrassing to statesmen if their countrymen insist on its observance.   If it were possible for the new Japan to live in company with her old-fashioned sentiments, they might be cited as the most clearly defined characteristics of her national policy.   But they will not survive; they have no element of " fitness " in this advanced age.   The instinct of self-preservation will compel her to take on the color of her "civilized " surroundings.

THE END.

TABLE OF CONTENTS.